Kugamoorthy Velauthamurty
Gopal Nakeeran

Computer aided animations for resonance in G.C.E. A/L Chemistry

AF138606

Kugamoorthy Velauthamurty
Gopal Nakeeran

Computer aided animations for resonance in G.C.E. A/L Chemistry

LAP LAMBERT Academic Publishing

Impressum / Imprint

Bibliografische Information der Deutschen Nationalbibliothek: Die Deutsche Nationalbibliothek verzeichnet diese Publikation in der Deutschen Nationalbibliografie; detaillierte bibliografische Daten sind im Internet über http://dnb.d-nb.de abrufbar.

Bibliographic information published by the Deutsche Nationalbibliothek: The Deutsche Nationalbibliothek lists this publication in the Deutsche Nationalbibliografie; detailed bibliographic data are available in the Internet at http://dnb.d-nb.de.

Coverbild / Cover image: www.ingimage.com

Verlag / Publisher:
LAP LAMBERT Academic Publishing
ist ein Imprint der / is a trademark of
OmniScriptum GmbH & Co. KG
Bahnhofstraße 28, 66111 Saarbrücken, Deutschland / Germany
Email: info@lap-publishing.com

Herstellung: siehe letzte Seite /
Printed at: see last page
ISBN: 978-3-659-81345-0

CONTENTS

IMPLEMENTATION OF COMPUTER AIDED ANIMATIONS TO TEACH RESONANCE IN G.C.E. ADVANCE LEVEL CHEMISTRY

Dr.K.Velauthamurty

The influence of computer aided animations against the traditional teaching method in teaching resonance to A/L chemistry students was investigated using two students groups namely control group and experimental group. As a pre-test, one hundred and twenty respondents who were randomly selected from the schools were tested on their existing understanding in resonance using evaluation sheets. They were ranked according to the scores they got. Students who were in even numbers at the ranking were absorbed into experimental group while others are considered as control group.

As the post-test, both groups were given same open-ended question papers. Mean, median, standard deviation, minimum marks and maximum marks of experimental group were calculated using MINITAB 14 statistical software and these attributes were compared that of control group. Experimental group which was prepared with the help of animated diagrams and self -study package showed higher performance to this post-test than the control group which was prepared only with the traditional teaching method, got to the same. P-values for the pairs of pre- test and post-test for both groups were separately calculated by hypothesis analysis and this value was obtained as 0.00 for both groups at 95% of confidence interval.

Respondents belong to experimental group were categorized into four groups such as no understanding, beginning, developing and mastery level according to the scores they achieved and it was compared that of control group. Number of students in the higher competence level or mastery level was remarkably increased in the experimental group compared with controlled group. Number of students at the no understanding and the beginning levels were observed very law in the experimental group when those were higher in the control group.

Chapter 1

Introduction

1.1 Background of the study

Teaching chemistry at the introductory level has made it obvious to the teachers that understanding resonance can be difficult and sometimes traumatic for students. Resonance is frequently a source of confusion when students are firstly exposed to it, and unfortunately, this feeling may linger even after repeated exposure. Understanding the three-dimensional aspects of molecules and the delocalization of electrons within the molecule is also difficult.

This study was aimed to determine the effects of visualization activities in the alternative teaching method on G.C.E (A/L) school students' conceptual understanding of 'Resonance' topic in organic chemistry and inorganic chemistry. In order to ascertain whether the particular topic is presented in accordance with the expected objectives set forth, evaluation of the existing teaching methodology is essential.

Understanding of methods which student used for the learning process, considered as an effective strategy for the learning. This requires that research into the learning process is made accessible. To facilitate the development of students' views of knowledge, students need to be supported at the appropriate level.

Chemistry is one of the most important branches of science; it enables learners to understand what happened around them. Because chemistry topics are generally related to or based on the structure of matter, chemistry proves a difficult subject for many students. This may sometimes repels learners from continuing with studies in chemistry. The complexity of chemistry has implications for the teaching of chemistry today.

The nature of chemistry concepts and the way the concepts are represented make chemistry difficult to learn. The methods by which students learn are potentially in

2

conflict with the nature of science, which, in turn, influences the methods by which teachers have traditionally taught.

In chemistry, resonance is a way of describing delocalized electrons within certain molecules or poly atomic ions where the bonding cannot be expressed by one single Lewis formula. Unlike in the past, now the students are expected to answer for many questions, which related to the resonance in the organic and inorganic compounds, in their own. However, in Advanced Level, many students are facing problem for drawing Lewis structure, to understand the delocalization techniques, identify the requirement where the resonance occurs and estimate relative stability of resonance structure.

Therefore, apart from the information in the syllabus they need an additional guide to promote their learning. Consider this situation; in this alternative approach, use of computer based Flash software, Chem. 3D pro, chem. draw and other visualization techniques are recommended to teach the topic effectively.

This report seeks to bring the general findings obtained from research in eight weeks for G.C.E (A/L) chemistry students in an attempt to suggest the key reasons for this difficulty. Suggestions are made on ways to minimize the problems based on understandings of attitudes and motivation as well as the psychological understandings of how learning takes place.

1.2 Aims and Objectives of the study

This study is aimed to determine the effects of visualization activities in the alternative teaching method on G.C.E (A/L) school students' conceptual understanding of 'Resonance' topic in organic chemistry and inorganic chemistry. The interactions and distinctions between them are important characteristics of chemistry learning and necessary for achievement in comprehensing chemical concepts. Therefore, if the students possess difficulties at one of these levels, it may influence the other. For example, if a student has difficulty on the resonance topic, it may influence in the reaction mechanisms of organic compounds, determination of the order of acidity and basicity of organic compounds and the geometric shapes of some inorganic compounds. Thus, determining and overcoming these difficulties are considered as the primary goal.

The main objective of this research work is to test how the alternative teaching strategies help to determine the understanding ability on improving the results in Advanced Level chemistry students. The aim of this paper is to determine and examine the specific advantages and disadvantages of the alternative teaching strategies.

In addition to this, the following objectives are desired to be achieved in carrying these activities.

- To make a clear understanding about resonance with the backgrounds of Lewis structure, electro negativity and delocalized nature of electrons.
- To prepare animations in order to make easy the resonance studies
- To apply chem Draw software to make easy the resonance studies
- To improve the students' knowledge and attitude in resonance
- To achieve a best output in Advanced Level chemistry.
- To contribute educational based research and discover by new methods that can help to improve chemistry teaching instructions.

This study is important, because, it contributes to increase the knowledge on teaching and learning. Furthermore, it helps to the Advanced Level chemistry students to understand more about the skills that is needed in learning resonance as fast as possible. Also, this study is significant to chemistry educators, because, it helps them to identify several issues that they can reflect on their teaching styles. The findings from this study also help them to improve their teaching skills by knowing which technique is benefit to students.

1.3 Review of Literature

Many students in secondary school and in the universities have many difficulties in understanding chemistry (Othman, Treagust & Chandrasegaran, 2008). For this reason, students develop scientifically unacceptable conceptions about many subjects or concepts in chemistry. Their knowledge of chemistry is therefore incomplete and incoherent (Kozma & Russell, 1997). Many students, in fact, merely memorize chemistry concepts without actually learning them (Othman et al 2008). This situation is an indication of why some students never come to like chemistry.

Conceptual understanding in chemistry is related to the ability to explain chemical phenomena through the use of macroscopic, molecular and symbolic levels of representation (Gabel, Samuel & Hunn, 1987). It is known that when relationships are formed between these three levels of representation, students understand and learn more in chemistry. In learning environments that include ICT, students are able to form successful relationships between the three levels of representation in chemistry and thus learn the subject in a more effective and meaningfully (Sanger, Phelps & Fienhold, 2000).

Individuals construct mental models to interpret phenomena and make sense of them (Johnson-Laird, 1983). A mental model is defined as an individual's personal description of a concept or event that has been impressed in that person's mind (Coll & Treagust, 2003). Through ICT, students rearrange their thoughts about chemical phenomena and processes and build meaningful mental models (Clark & Jorde, 2004). ICT provides students the opportunity of improving their conceptual understanding and forming mental models of high quality. The role of ICT in student-centered education is to provide tools whereby the student's comprehension ability can be increased.

Teachers should be provided with scientific explanations as to what the teacher's responsibility is and should be, within the framework of constructivist teaching that encompasses ICT. Furthermore, it will also be very important to enhance teachers' knowledge about how exactly to benefit from technological tools in the teaching environment. Teacher education should not only include technical information as to how to use the technology but should also cover how to choose the right methods and strategies to be used in the teaching environment where technological tools are employed. Teachers should be informed about the benefits of technological tools which can offer students, when used in the classroom. For example, some chemical reactions may constitute a serious risk for students if carried out on their own. Instead of having students work on such reactions, possible risks might be avoided by using ICT to demonstrate (Clark et al., 2004).

There are significant research studies in science which focus on students' alternative conceptions which could influence the learning process in their mind. As a result, it becomes challenges for the teachers to recognize and guide these students' alternative conceptions. Moreover, it is common perceptions that the teachers face the difficulties to explain the concepts in chemistry. Students could reject or accept, or assimilate the concepts or ideas which are given by the teacher, because they already have their own concepts which recognize as "prior knowledge". This prior knowledge could be strongly hold by students and it is very difficult to be changed (Clark et al., 2004).

There are many terms that are used regarding to alternative conceptions, such as misconceptions, intuitive ideas, interpretive frameworks, children's science, etc. Based on research study on "An Inventory for Alternate Conceptions among First-Semester General Chemistry Students", Mulford and Robinson (2002) used the diagnostic instrument which develops by Treagust (1988) (as cited in Mulford & Robinson, 2002). They found that one of students' alternate conceptions in chemistry on the conservation of mass, molecules, and atoms during a chemical reaction is "the total number of molecules is also conserved in a chemical reaction". Moreover, the other examples of students' alternative conceptions in chemical bonding based on octet rule are:"

1) Atom "need" filled shell,

2) A covalent bond holds atom together the bond is sharing electrons,

3) Molecule forms from isolated atoms" (Robinson et al, 1993,).

These examples are only few students' alternative conceptions which influence the students to understand the chemistry concepts. Teachers could use the information of students' alternative conceptions to choose the best teaching strategies to guide these conceptions. As a result, students could engage with chemistry concepts through their existing ideas. Therefore, it is important for teacher to recognize students' conceptions before introducing the new topics in chemistry.

Mathematics curricula based on multiple representations have been found to provide teachers with a greater variety of instructional and assessment approaches than do traditional curricula. Similar findings have also been reported in chemistry classrooms (Russell et al., 1997). Participants in the collaborations described here devised a set of characteristics that should be considered when introducing molecular visualizations into the tertiary general chemistry curriculum. These include providing opportunities for practice and feedback on learning, appropriate annotation of visualizations, gradual introduction of conventions and structural complexity, comparative presentation of related visualizations, the use of animations, provisions for appropriate interactions with the visualizations, and instructional materials designed for concept mastery and inquiry.

Interactive computer-generated molecular visualizations provide many opportunities for instructional innovation. They are being used not only to encourage students to think about chemistry in terms of molecules, models, and symbols, but also to provide opportunities for students to become more independent learners. In addition, interactive visualizations can help instructors interest and motivate students with a variety of learning styles. To achieve these goals, more activities incorporating visualizations that are easy for students to use and interpret will need to be developed. Many topics in chemistry require learners to understand structures in three dimensions, changes over time, and causality. Molecular animations can be powerful tools for learning these dynamic and three-dimensional chemistry concepts. However, a problem with the use of animations for teaching molecular structure and dynamics is that merely viewing a visualization may lead to learning at a lower level than would drawing or building a molecular structure. Consequently, it may be beneficial for computer-based visualizations to provide opportunities for students actively to explore concepts (Sanger et al., 2000).

He studied how the students used molecular models and 3D computer visualizations to develop an understanding of stereoisomerism. He also obtained a model kit and an organic chemistry textbook in order to learn the stereochemistry concepts taught in the course.

1.4 Implication of Educational Research

This educational research involves computer based Flash software, Chem. draw and other visualization techniques. Visualization tools and high performance computing have changed the nature of chemistry research and have the promise to transform chemistry instruction. Visualization tools to be useful in education, students must be able to interpret the images they produce. These images are used to convey complex, subtle molecular interactions and dynamics that are difficult to describe in words. Using these tools requires the ability to identify and make use of complex visualizations of molecular structures.

Some chemical phenomena are not obvious without the use of visualizations. It also helps to the students learn about molecular structure and dynamics, electron arrangement in molecules, a strategy for writing Lewis structure and the delocalization of electron pair in molecules known as Resonance in molecules. At the beginning, chemistry students can now be exposed to a wide array of molecular visualizations: structural formulas, line drawings, physical models, and a variety of dynamic three-dimensional computer-generated molecular models.

Since students are always exposed to traditional teaching method where there is a use of board and marker (chalk) and almost all students have not been exposed to such a class room that occupies computer aided facilities (e.g. animations), at the beginning, having a special attention at them is important to understand the electron flows at newly introduced animated digammas. For example, repeating the showings for the electron flows and making sure their self and or group studies apart from the class room. Identifying absentees for the lessons and making up them for next lesson are also important. Hence, this study facilitates teachers and students to have effective teaching learning process at the classroom level in the discipline of resonance in Chemistry.

Chapter 2
Methodology

2.1 Sample selection

This study used mixed method approach by considering its advantages. Thus, data from questionnaires and interviews were collected in this study.

One hundred and twenty students who are following chemistry as a subject in their G.C.E. Advanced Level and Grade Eleven chemistry teachers were randomly selected from Vidyananda College, Mullaitivu Maha Vidyalaya, Mallavi Central College and Oddusuddan Maha Vidyalaya in Mullaitivu district, Sri Lanka.

School	Number of students
Mallavi Central College , Sri Lanka	25
Mullaitivu Maha Vidyalaya , Sri Lanka	19
Vidyananda College , Sri Lanka	76

Table 2.1 Sample selection from school students

2.2 Collection of data

Primary data collection was performed from the students and teachers as well by giving two, separate, questionnaires. Students were asked many questions about their thinking about resonance. Replies of respondents for the questionnaires which were made in a qualitative way were quantitatively analyzed (mixed method). Then, as a pre-test, the students were examined to test their existing knowledge in resonance. The question sheet for students was created while covering the following areas:

- Lewis structure;

- Bond order, bond length and bond strength;

- The requirement where the resonance occurs;

- The features of resonance; and

- Rules to estimate relative stability of resonance structure.

The examination sheet contained items of various varieties such as short answer, multiple choices and structured essay questions. The pre-test was consisted of 12 questions and each question contained sub questions. Then, the students were ranked in ascending order according to the scores they got to the pre-test. After the ranking, they were equally divided into two groups, namely control group and experimental group. Students who were in odd numbers at ranking were absorbed into experimental group and the other students were categorized as control group. Mean, median, standard deviation, maximum marks and minimum marks for both groups were calculated. The control group was taught by traditional teaching method, "chalk and talk" whereas the experimental group was treated with alternative teaching methods, specially using power point presentation and animated diagrams. Responses of students for the previously issued questionnaire were taken in mind during the alternative teaching method. A power point based study pack that emphasizes the students' self-learning was also offered to the experimental group. Students were grouped in order to include three of them for each group for the above self-learning system. The lesson plan was 10 periods. After the completion of treatments, as post-test, it was given a question paper regarding resonance for both groups and the scores they received in each group were statistically analyzed using MINITAB 14 statistical software. Mean, maximum marks, and minimum marks of each group were tabulated.

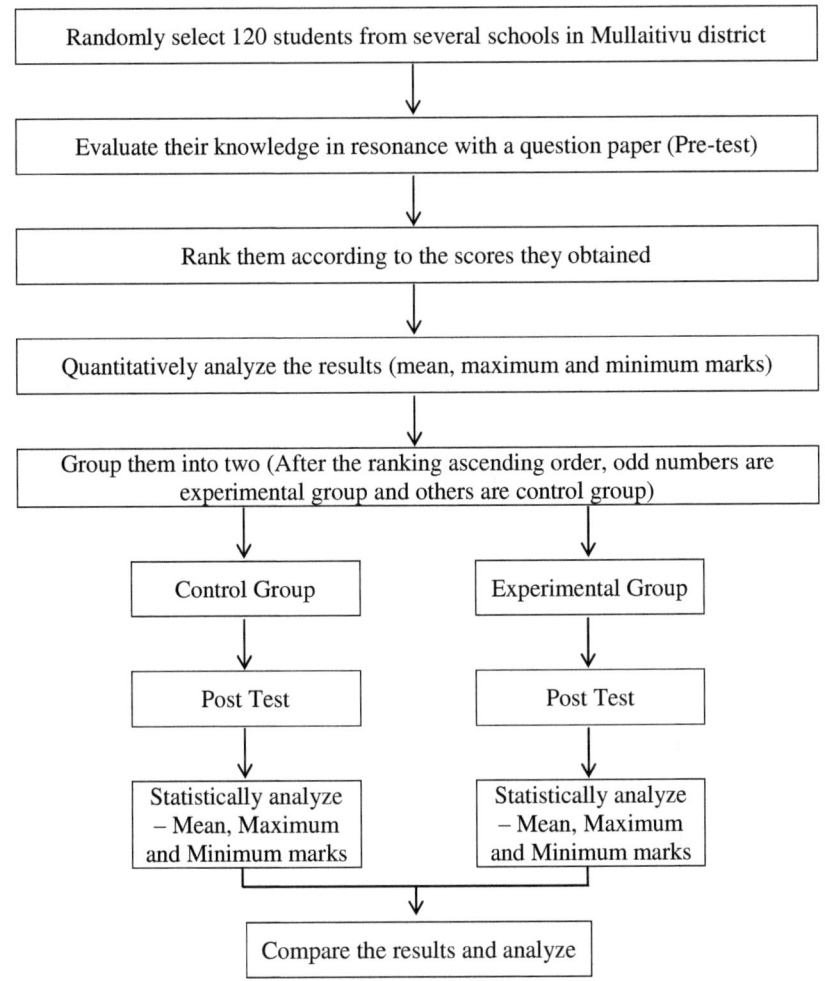

| Randomly select 120 students from several schools in Mullaitivu district |

| Evaluate their knowledge in resonance with a question paper (Pre-test) |

| Rank them according to the scores they obtained |

| Quantitatively analyze the results (mean, maximum and minimum marks) |

| Group them into two (After the ranking ascending order, odd numbers are experimental group and others are control group) |

| Control Group | Experimental Group |

| Post Test | Post Test |

| Statistically analyze – Mean, Maximum and Minimum marks | Statistically analyze – Mean, Maximum and Minimum marks |

| Compare the results and analyze |

Table 2.2 Pre-test and post-test plan for students

Then the control group and experimental group were separately analyzed, using paired-t test in hypothesis for pre-test and post test. P-values for both were found out.

On the other hand, selected chemistry teachers were requested to report their responses in a qualitative manner at the questionnaire and their responses were quantitatively analyzed in order to bring out the difficulties they face in teaching the topic "resonance" in schools.

2.3 Materials

Pre Test - The pre-test, composed of 12 open-ended questions, each having sub-questions, was developed using the teachers' guide and text books. Questions in the test were mainly at knowledge/comprehension levels according to the resonance topic. The validity of the test was achieved by consulting five chemistry teachers. With respect to the reliability, the pre-test was administered to a group of hundred and twenty students who will take G.C.E (A/L) exam in the next year. A sample question can be seen in Appendix 1.

2.3.1 Alternative teaching materials

Teaching instruction is based on a series of very small steps, called slides. Each slide contains some information, animation diagram that shows electron movement and some important questions. Students should write answers in a blank sheet for the questions. Answers were analyzed by me. Then the students write the correct answer before going to the next slide. If the student's answer was correct, it is positively reinforced by progress to the next slide; if not, the student immediately sees the correct answer. Each slide may introduce either a new idea or repeat material covered earlier. The lessons start from the student's initial knowledge and in small steps proceeds to a final learning goal. Because of active student participation, small steps, immediate feedback and reinforcement, alternative approach can be very effective. All students work through the same sequence. Every question is followed by another slide which contains the answer. In this study linear sequencing was employed. A total of fifty slides were prepared, covering all resonance concepts and principles at introductory level. A sample of slide can be seen in Figure 2.

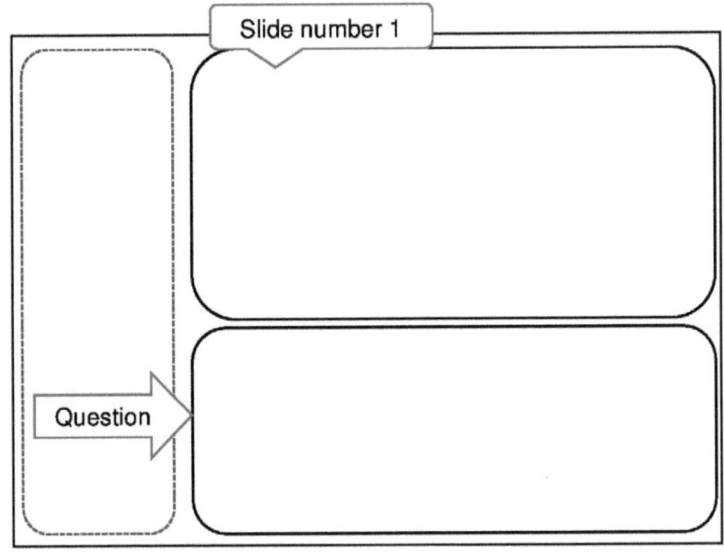

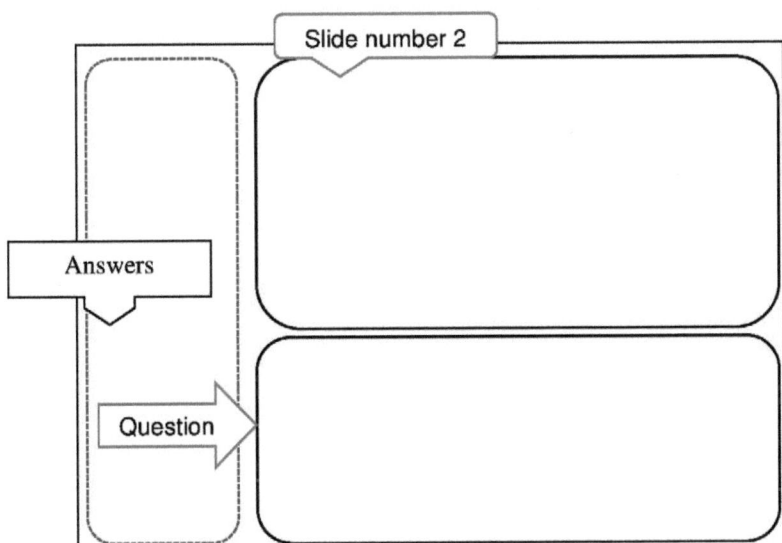

Figure 2.3.1 A sample slide of alternative teaching approach

Figure 2.3.2 Self learning of students in alternative method

Figure 2.3.3 Self learning of students in alternative method

2.4 Preparation of the teaching material

This educational research involves computer based Power point, chem. draw, colorful diagrams, animated diagrams that show the movement of free electrons within the compound and other visualization techniques. Here I used the PowerPoint program and chem. draw software to design the molecular structure.

For the convenience of the research, the topic was divided into following sub topics. They are;

- Lewis structure;
- Fundamental requirement of a Lewis structure;
- The strategy for writing Lewis structure;
- Introduction of resonance;
    Here introduces the bond order, bond length and bond strength;
- The requirement where the resonance occurs;
- The features of resonance; and
- Rules to estimate relative stability of resonance structure.

Lewis structure

Resonance structure is an average of two or more Lewis structures which differ only in the position of their electrons. Therefore, before understanding the resonance phenomena the student must know to draw the Lewis structure. A Lewis structure is a representation of covalent bonding in a covalent molecule using Lewis dot symbols in which shared electron pairs are shown either as lines or pairs of dots and lone pairs are shown as pairs of dots or crosses on individual atoms. Only valence electrons are drawn in a Lewis structure.

Example: for H_2O molecule
- Lewis symbol for oxygen
- Lewis symbol for hydrogen
- Lewis structure for water

The Lewis representation of some elements

Li $\overset{\cdot}{\underset{\cdot}{Be}}$ $\cdot \overset{\cdot\cdot}{N} \cdot$ $\cdot \overset{\cdot}{\underset{\cdot}{C}} \cdot$

$: \overset{\cdot\cdot}{\underset{\cdot\cdot}{Ne}} :$ $\cdot \overset{\cdot\cdot}{\underset{\cdot\cdot}{O}} \cdot$ $\cdot \overset{\cdot}{\underset{\cdot}{B}}$

The Lewis representation for some molecules

H₂   H−H

HCl $H - \overset{\cdot\cdot}{\underset{\cdot\cdot}{Cl}} :$

O₂ $: \overset{\cdot}{\underset{\cdot}{O}} = \overset{\cdot}{\underset{\cdot}{O}} :$

NH₃   H−$\overset{\cdot\cdot}{N}$−H
             |
             H

N₂ $: N \equiv N :$

Fundamental requirement of a Lewis structure

1. All the valence electrons of the atoms in a Lewis structure must be shown in the structure.
2. Usually, each atom in a Lewis structure acquires an electron configuration with an outer shell octet. Hydrogen, however, is limited to an outer shell duet.
3. Usually, all electrons in a Lewis structure are paired.
4. Sometimes, double and triple bonds are included to represent a Lewis structure. Multiple covalent bonds are readily formed by C, N, O, P and S.
5. Start with the correct skeleton structure.

Strategies for writing Lewis structures

I. Write a skeleton structure of the molecule by joining the atoms in the structure by single covalent bonds. When there is more than one possibility for the skeleton, atom that is least electronegative is the central atom. Hydrogen always usually occupies the terminal position in the Lewis structure.

Example- Consider the SO_3^{2-} ion

$$
\begin{array}{c}
O \\
\uparrow \\
| \\
O \!\!-\!\! S \!\!-\!\! O
\end{array}
\quad \text{(S least electronegative element)}
$$

II. Establish the total number of valence electrons in the skeleton structure. For polyatomic ions, add extra electron for each negative charge and subtract one electron for each positive charge.

$$
\begin{array}{rcl}
S & = & 6 \\
3 \times O & = & \underline{18} \\
& & 24 \\
\text{Negative charge} & & \underline{+2} \\
& & \underline{26}
\end{array}
$$

III. For each single bond in the skeleton structure, subtract two electrons from the total number of valence electrons obtained in (II). Convert this number to electron pairs by dividing by 2.

$$26 - 6 = 20/2 = 10 \text{ electron pairs}$$

IV. Distribute these electron pairs first around the terminal atoms in the skeleton such that they acquire the noble gas configuration and then to the, extent possible, around the central atom(s).

$$:\ddot{\text{O}}-\ddot{\text{S}}-\ddot{\text{O}}:$$
$$| \atop :\ddot{\text{O}}:$$

V. If the central atom lacks an octet, then convert single bonds into double bonds, failing double bonds to triple bonds (i.e. multiple covalent bonds) by shifting lone pairs from the terminal atoms.

$$:\ddot{\text{O}}=\ddot{\text{S}}-\ddot{\text{O}}:$$
$$|^{-} \atop :\ddot{\text{O}}:$$

VI. Assign the formal charges on atoms.

$$\begin{array}{l} \text{Formal} \\ \text{charge} \\ \text{(FC) of} \\ \text{an atom} \\ \text{in Lewis} \\ \text{structure} \end{array} = \begin{array}{l} \text{Number} \\ \text{of} \\ \text{valence} \\ \text{electrons} \end{array} - \frac{1}{2} \left\{ \begin{array}{l} \text{Number} \\ \text{of} \\ \text{bonding} \\ \text{electrons} \\ \text{around} \\ \text{the atom} \end{array} \right\} - \begin{array}{l} \text{Number} \\ \text{of lone} \\ \text{pairs on} \\ \text{the} \\ \text{atom} \end{array}$$

Formal charges present should be as small as possible to give stable Lewis structure. Since Sulphur belongs to the third row in the periodic table, it can take up more than 8 electrons around it.

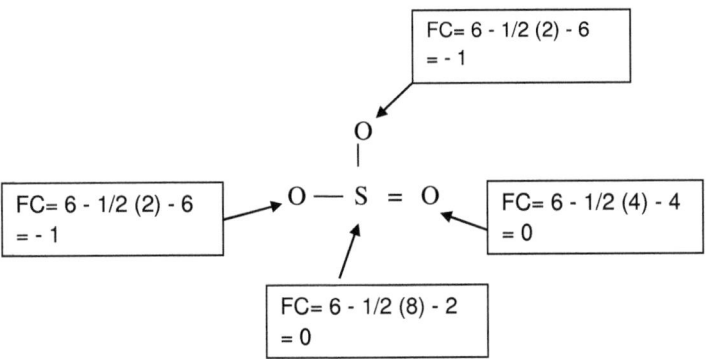

$$FC = 6 - 1/2\,(2) - 6$$
$$= -1$$

$$FC = 6 - 1/2\,(2) - 6$$
$$= -1$$

$$FC = 6 - 1/2\,(4) - 4$$
$$= 0$$

$$FC = 6 - 1/2\,(8) - 2$$
$$= 0$$

O — S = O

Figure 2.2 Lewis structure with Formal Charge calculations

Introduction of resonance

✍ Definition: The continuous delocalization of electron pair in an ion/ molecules is known as resonance.

✍ Simple example of O_3 can be considered to highlight the resonance phenomena.

✍ The Lewis structure of O_3 molecule can be written using the rules laid down in the previous lesson as follows.

✍ On the basis of above structure there must be two different O – O bond distances, one of them similar to O – O single bond of 147 pm in H_2O_2 and other similar to O – O double bond distance of 121 pm in O_2 molecule. The fact that observed O – O bond distance (128 pm) is same in O_3 somewhat longer than double bond and shorter than single bond distance it reveals it must have resonance phenomena.

✍ The above two Lewis structures are resonance structure.

✍ The resonance structures are defined as contributing structure or resonance contributors where they are differ only in the arrangement or distribution of electron.

✍ The resonance structures or resonance form a hypothetical or imaginary in the case of O_3 two equivalent Lewis structure contributes equally to the resonance hybride.

✍ Resonance hybrid is the real structure.

$$-1/2 \quad \overset{\overset{\displaystyle ..}{\displaystyle +}}{O} \quad -1/2$$
$$\underset{\underset{\displaystyle ..}{}}{:\ddot{O}} \diagdown\ \diagdown\ \underset{\underset{\displaystyle ..}{}}{\ddot{O}:}$$

✍ Bond order of O_3 molecules is 1.5 that is calculated as follows

$$\text{B.O} = \frac{\Sigma \text{ Number of bonds between two considered atoms}}{\text{No. of Resonance structures}}$$

$$\text{B.O} = \frac{2+1}{2}$$

$$= 1.5$$

Bond order is the number of bond present between considered two atoms. Therefore, in resonance hybrid of O_3 molecule, are can consider 1.5 $O - O$ bonds are present, between two O atoms.

When Bond order increases bond length decreases. Bond strength increases. Therefore, we can say the bond strength of $O - O$ bond of O_3 is between the bond strength of $O = O$ and $O - O$.

The instances where the Resonance occurs

1. When conjugated double bond is present

$$H_2\bar{C}-\underset{H}{C}=\underset{H}{C}-\overset{+}{C}H_2$$

$$\updownarrow$$

$$H_2C=\underset{H}{C}-\underset{H}{C}=CH_2$$

$$\updownarrow$$

$$H_2\overset{+}{C}-\underset{H}{C}=\underset{H}{C}-\bar{C}H_2$$

2. When charged atom/group directly attached with double bonded atom

$$H_2C=\underset{H}{C}-\overset{+}{C}H_2 \quad \longleftrightarrow \quad H_2\overset{+}{C}-\underset{H}{C}=CH_2$$

3. When an atom having lone pair directly attached with double bonded atom.

22

The features of resonance

1. The compound stability is increases when the resonance occurs.

$$H_2C=C-C=CH_2 \overset{H\ H}{} \quad > \quad H_2C=C-C\overset{H_2}{}C=CH_2 \overset{H}{}\overset{}{}\overset{H}{}$$

2. The ion or molecules stability is increases when the number of resonance structure increase.

$$ClO_3^- < ClO_4^-$$

3. Net Charge is not change when the resonance occurs.

$$:\ddot{O}=C=\ddot{O}: \longleftrightarrow :\overset{-}{\ddot{O}}-C\equiv\overset{+}{O}: \longleftrightarrow :\overset{+}{O}\equiv C-\overset{-}{\ddot{O}}:$$

4. The ion or molecule's energy of the resonance structures is approximately equal.

Rules to estimate relative stability of resonance structure

1. The most stable resonance structure has the least formal charge distribution with the greater no of covalent bonds and it contributes mostly to the resonance hybrid.

2. Structures in which adjacent atoms have formal charges of the same sign tend to unstable.

3. When opposite charges are placed on atoms positive charges should fall on electro positive atom and negative charge on electro negative atom.

2.5 Implementation of the research plan

Treatment was completed in a total of ten periods. Each control group was taught by the teachers of the respective schools. The students were guided, as in previous years, by using a traditional 'chalk and talk' method with conventional teaching approach. Here the students took chemistry tutoring without the help of visual aids and diagrams.

The experimental group had sixty students. It was divided into twenty sub groups. Then I made a copy of animation diagrams with the help of lab technician in 20 computers in the respective schools. These slides were prepared by using the computer based power point program. When a student completed the given slide, he/she was presented with the next slide. The rate of progress depended on the group of students. In this way, the speed of the students ranged from 5 to 7 slides in each class for 1 period. Here the students took chemistry tutoring with using the visual aids, colorful diagrams, animation diagrams that show the movement of free electrons within the compound, the descriptions with Over Head Projector (OHP) and finally gave an opportunity to write notes on their own.

I took mainly a tutoring role rather than that of an instructor. I guided the students in using the slides and helped them in places where students needed explanations and extra help. In this way, all the students were kept active, and they were involved in the learning process. During this process neither additional information was given nor extra problems solved beyond those which were given in the slides. However, students were free to get information outside of class hours. Each student had the chance of learning about resonance at his/her own pace.

When the treatment was completed, both the control and experimental groups were given the post-test. The effectiveness of the alternative approach on the experimental group was evaluated using a statistical method. Moreover, students' views about the alternative approach were gathered from the experimental group. For this purpose, students were given a blank sheet and they were asked to write their views about the treatment. They were asked not to write their names on the sheets in order to ensure confidentiality.

Use of multimedia projector and other audio visual aids in class room teaching needs careful time management. While prominence is given to enhance perception and meaningful understanding of the lesson, it is equally important to consider time factor too.

Chapter 3

Results and analysis

3.1 Results

Quantitatively analyzed results for the separate questionnaires which were issued to thirty students and eleven chemistry teachers before beginning the pre-test are shown in the following tables.

#	Questions	Total Number of Respondents	Yes (%)	No (%)	No Response (%)
1	Do you agree to memorize resonance concepts and mechanisms without actual understanding them?	30	7 (23%)	23 (77 %)	
2	Do you easily understand the concept and mechanisms in resonance without applying any technical tools?	30	9 (30 %)	21 (70 %)	
3	Do you prefer to use technical tools in your chemistry class room to make easy your understanding in these subjects?	30	26 (87 %)	4 (13 %)	
4	Do you refer Books, Journals or Internet before your teacher start the particular lesson?	30	3 (10 %)	27 (90 %)	

#		Total Number Of respondents		
5	Are you most preferred to book study or to use visualized aids such as animations in your study?	30	22 (73 %)	8 (27 %)
6	Have your teachers used any animations in your chemistry class room?	30	5 (17 %)	25 (83 %)
7	Do you prefer to study alone?	30	17 (57 %)	13 (43 %)
8	Do you prefer to have a "group discussion"?	30	23 (77 %)	7 (23 %)

Table 3.1: Summary of the outcome for the questionnaire issued to students

#	Questions	Total Number Of respondents	Yes (%)	No (%)	No Response (%)
1	Do you think that the resonance is the important topic in G.C.E. A/L chemistry?	11	10 (91 %)	-	1 (9 %)
2	Have you realized that students are struggling in drawing the mechanisms of resonance	11	7 (64 %)	3 (27 %)	1 (9 %)
3	Do you feel difficult to teach the resonance mechanisms to the students?	11	7 (64 %)	4 (36 %)	

4	Do you think the need of alternative teaching material is important to teach the resonance topic?	11	9 (82 %)	2 (18 %)	
5	Do you think that the introduction of animation diagrams to show the particular resonance phenomena will have favorable impact on understanding the topic, "resonance"?	11	10 (91 %)	1 (9 %)	
6	Do you use computer facilities to teach the topic – 'Resonance' in your class room?	11	2 (18 %)	9 (82 %)	
7	Do you think the power point based presentation with the use of Over Head Projector (OHP) is important to teach the resonance topic?	11	7 (64 %)	4 (36 %)	
8	Do you agree to use such technical tools in your classroom?	11	8 (73 %)	3 (27 %)	

Table 3.2: Summary of the outcome for the questionnaire for the teachers

3.2 Quantitative analysis for the questionnaire issued for students

A total of 30 students were asked to report during the study period and the outcome is presented in Table 3.1. The study revealed that:

- 77 % of the respondents did not agree to memorize resonance mechanisms without actual understanding
- Only 30 % of selected students could be able to understand resonance mechanism in a conventional class room while rest of the respondents faced difficult on it.
- 87 % of the students supported to the question "Do you prefer to use technical tools in your chemistry class room to make easy your understanding in these subjects?". Only 13 % of respondents disagreed to this question.
- Only 10 % of students referred books, journals and internet whereas 90 % of students never refer any books, journals or internet before they learn the subject at school.
- 73 % of the students preferred to use visualized aids such as animation during their resonance study while only 17 % of the students positively answered to the question "Have your teachers used any animations in your chemistry sessions?"
- 57 % of the respondents preferred to study first, alone. But 77 % the total students wanted to have a group discussion after finishing their individual study.

3.3 Quantitative analysis for the questionnaire issued for teachers

- A total of 11 teachers were interviewed during the study period and the outcome is presented in Table 3.2.

- The study revealed that a 91 % of the selected respondents agreed that the resonance is the important topic in G.C.E. A/L chemistry. Remaining 9 % of them did not responses.

- 64 % of the teachers agreed to the question, "Have you realized that students are struggling in drawing the mechanisms of resonance". 27 % of teachers did not agree with this status whereas rest of the respondents (9 %) shown no response to the statement.

- 64 % of the teachers replied "yes" to the question, "Do you feel difficult in teaching the resonance mechanisms?" And 82 % of total teachers positively answered to the question, "Do you think the need of alternative teaching material is important to teach the resonance topic?. In contrast, 36 % of the teachers did not feel difficult to teach the topic to the students and 18 % of them out of 100 % did not think that the need of alternative teaching material is important to teach the topic "resonance".

- 91 % of the respondents replied that they think that the introduction of animation diagrams to show the particular resonance phenomena will have favorable impact on understanding the topic, "resonance". Rest of them retrieved the statement.

- Only 18 % of the teachers have used computer facilities to teach the topic 'resonance' in the class room.

- 73 % of the teachers agreed to use technical tools such as animations in the classrooms and the remaining 27 % disagreed with this statement.

Name of the School	Number of students	
	Control group	Experimental group
Mallavy Central College, Sri Lanka	15	10
Mullaitivu Maha Vidyalaya, Sri Lanka	08	11
Vidyananda College, Sri Lanka	36	40

Table 3.3: Numbers of students for post -test according to school wise

3.4 Scores of the students for the Pre – test and Post -test

#	Experimental Group	
	Outcomes for the Pre- test	Outcomes for the Post -test
1	10	47
2	14	68
3	14	52
4	15	66
5	15	74
6	18	76
7	18	58
8	19	58
9	19	66
10	20	50
11	20	59
12	20	68
13	20	52
14	20	57
15	25	52
16	25	66
17	28	57
18	29	62
19	30	49
20	30	54
21	30	62
22	30	65
23	32	50
24	35	84
25	35	90
26	35	73
27	38	70
28	39	65
29	40	59
30	40	80
31	40	96
32	40	72
33	40	84
34	42	58
35	42	73
36	43	82

37	44	87
38	45	85
39	45	59
40	45	89
41	46	96
42	46	92
43	46	87
44	47	68
45	48	72
46	48	77
47	49	73
48	50	86
49	50	90
50	50	87
51	50	75
52	50	86
53	51	93
54	52	90
55	54	88
56	54	83
57	54	81
58	55	72
59	55	48
60	58	81

Table 3.4. Summary of the pre-test and post-test scores, out of 100, for the experimental group.

Control Group		
#	Pre-test	Post-test
1	12	37
2	14	33
3	15	28
4	15	34
5	15	40
6	18	33
7	19	42
8	19	32
9	19	48
10	20	35
11	20	42
12	20	39
13	20	47
14	24	44
15	25	46
16	28	42
17	28	50
18	29	58
19	30	38
20	30	42
21	30	56
22	32	47
23	34	50
24	35	69
25	35	25
26	37	47
27	38	53
28	39	51
29	40	27
30	40	34
31	40	56
32	40	32
33	42	54
34	42	48
35	42	47
36	44	53
37	45	50
38	45	57
39	45	52

40	45	40
41	46	66
42	46	47
43	47	63
44	47	67
45	48	45
46	48	58
47	49	66
48	50	56
49	50	63
50	50	68
51	50	71
52	50	77
53	52	56
54	52	69
55	54	64
56	54	58
57	54	53
58	55	67
59	55	73
60	60	38

Table 3.5 Summary of the pre-test and post-test scores, out of 100, for the control group

3.5 Analysis according to the results

Descriptive Statistics: Experimental Group – Pre test, Control group-Pre test

Variable	Mean	St.Dev	Minimum	Median	Maximum
Experimental- Pr	36.70	13.38	10.00	40.00	58.00
Control group-Pr	37.12	13.23	12.00	40.00	60.00

Here the means for the pre-tests for both experimental group and control group are approximately equal (36.70 and 37.12 respectively). Marks of the 120 students were arranged in ascending order. Then, the marks which are related to odd numbers were considered as experimental group whereas even numbers were considered as control group. Using the above method, the two groups employed in this research project was arranged in order to intelligence wise equally distribute.

That is why both categories have approximately equal means.

Descriptive Statistics: Experimental Group - Pre test, Experimental - Post test

Variable	Mean	St.Dev	Minimum	Median	Maximum
Experimental- Pr	36.70	13.38	10.00	40.00	58.00
Experimental – P	71.65	14.12	47.00	72.00	96.00

In the experimental group, mean for the pre-test is 36.70 and the mean for the post-test is 71.65. It shows a significant increasing in the mean. Moreover, the minimum marks obtained during the pre-test is 10 and the minimum marks achieved during the post-test is 47. It also shows improvement among the selected student. Highest marks given during the pre-test is 58 and it is 96 for the post-test group. A major reason for these improvements may be because of the introduction and implementation of using technical tools in order to make easy understanding the resonance mechanisms while it also attract or motivate the students.

Descriptive Statistics: Control Group -Pre test, Control group - Post test

Variable	Mean	St.Dev	Minimum	Median	Maximum
Control group-Pre	37.12	13.23	12.00	40.00	60.00
Control group - post	49.72	12.73	25.00	49.00	77.00

Mean for the post-test in the control group has increased than the mean for the pre-test. But the range between two values are smaller than the difference in the means that obtained at experimental group. Besides, the maximum marks achieved by the students is 96 in the experimental group whereas the maximum marks for the control group is 77. It can be observed that the attributes for the experimental group are always better than the results of the control group.

Paired T-Test and CI: Experimental- Pre test, Experimental Group - Post test

Variable	N	Mean	St.Dev	SE Mean
Experimental- Pr	60	36.7000	13.3776	1.7270
Experimental - P	60	71.6500	14.1155	1.8223
Difference	60	-34.9500	12.24533	1.5809

Paired T for Experimental- Pre test - Experimental - Post test

95% CI for mean difference: (-38.1133, -31.7867)

T-Test of mean difference = 0 (vs not = 0): T-Value = -22.11 P-Value = 0.000

Null hypothesis has been applied for the experimental group.

Null hypothesis (H_O):- There is no statistically significant difference between the mean scores taught through alternative teaching approach

Alternative hypothesis (H_A):- There is a statistically significant difference between the mean scores taught through alternative teaching

The null hypothesis is rejected when the *p*-value is less than the predetermined significance level which is often 0.05 (at 95% confidence interval).

Here the *p*-value 0.00 which is less than 0.05. Therefore, null hypothesis (Ho) is rejected and the alternative hypothesis is accepted. Therefore, there is a statistically significant difference between the mean scores taught through alternative teaching approach.

Paired T-Test and CI: Control group-Pre test, Control Group - Post test

Variable	N	Mean	St.Dev	SE Mean
Control group-Pr	60	37.1167	13.2321	1.7083
Control group	-60	49.7167	12.7307	1.6435
Difference	60	-12.6000	11.0564	1.4274

Paired T for Control group-Pre test - Control group - Post test

95% CI for mean difference: (-15.4562, -9.7438)

T-Test of mean difference = 0 (vs not = 0): T-Value = -8.83 P-Value = 0.000

Again hypothesis has been applied:

Null hypothesis (H_O):- There is no statistically significant difference between mean scores taught through conventional teaching approaches.

Alternative hypothesis (H$_A$):- There is a statistically significant difference between the mean scores taught through the conventional teaching approaches.

The null hypothesis is rejected when the *p*-value is less than the predetermined significance level which is often 0.05 (at 95% confidence interval).

Here the *p*-value 0.00 which is less than 0.05. Therefore, null hypothesis (Ho) is rejected and the alternative hypothesis is accepted. Therefore, There is a statistically significant difference between the mean scores taught through conventional teaching approach.

Attribute	Post-test for experimental group	Post-test for control group
Mean	71.65	49.72
Maximum marks achieved by the student	96	77
Minimum marks obtained by the student	47	25

Table 3.6. Comparison of attributes within two groups in post-test.

Although there is significant difference in the conventional teaching approach, mean of the experimental group in the post test and the maximum marks achieved during the post test in experimental group are higher than that of control group. Moreover, the minimum marks obtained the experimental group is higher than the minimum marks obtained during the control group test as shown in the following table.

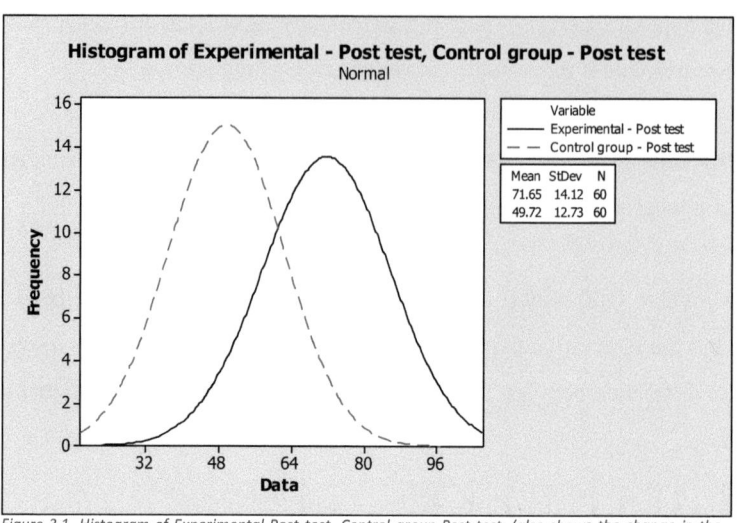

Figure 3.1. Histogram of Experimental-Post test, Control group-Post test. (also shows the change in the mean in the post-tests of experimental group and control group).

Range of the marks	Competence level of the student	Percentage of the students Belong to each level for Control group	Percentage of the students Belong to each level for Experimental group
01-29	No understanding	5 %	0 %
30-49	Beginning	45 %	5 %
50-74	Developing	48%	53 %
75-100	Mastery	2 %	42 %

Table 3.7 Student competency level of post test

When compare the post test marks of control group and experimental group with competence level, mastery level of the experimental group is higher than the control group.

By inspection of the above table, it is obvious that the performance of the experimental group is better than the controlled group. For example, according to the last column in the Table 3.6, 95% of students have scored over 50 marks whereas the 3rd column in the table indicates that only 50 % of students have scored over 50. Similarly, number of students in the higher competence level has remarkably increased in the experimental group compared with controlled group. (This general improvement can be reinstated by a statistical analysis). These results can be represented by the following bar-chart diagram.

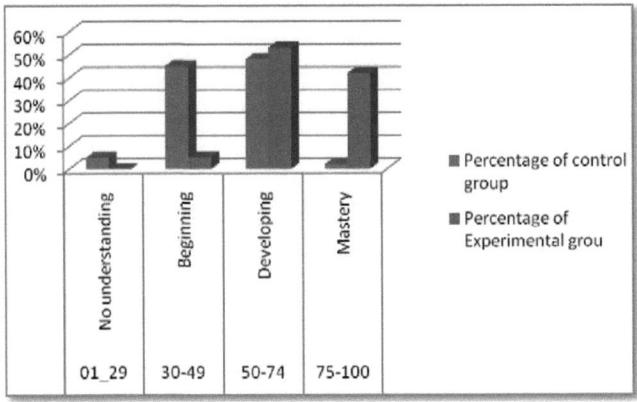

Figure 3.2 Student competency levels of post test control and experimental group

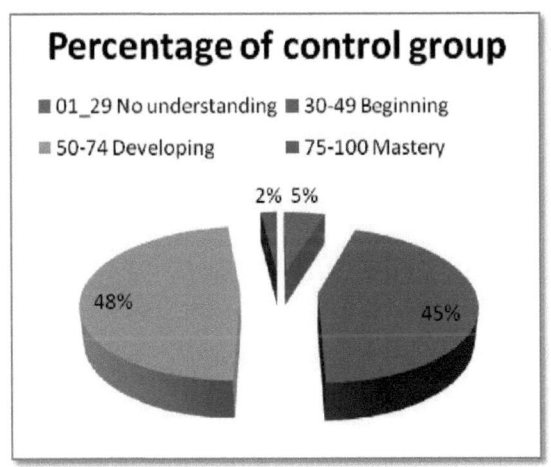

Figure 3.3 Percentage of Student competency level of post test in control group

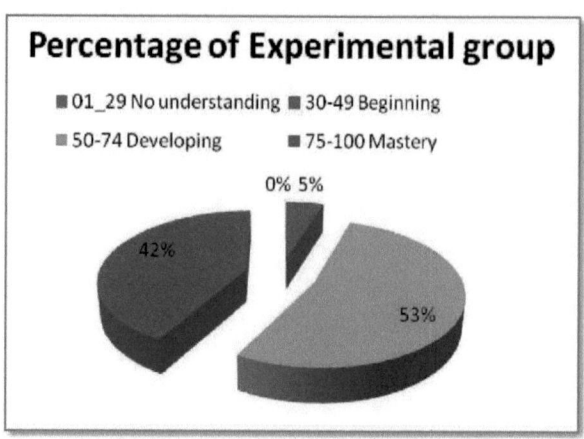

Percentage of Experimental group

■ 01_29 No understanding ■ 30-49 Beginning
■ 50-74 Developing ■ 75-100 Mastery

0% 5%

42%

53%

Figure 3.4 Percentage of Student competency level of post test in Experimental group

Chapter 4

Discussion and conclusion

4.1 Discussion

Questionnaires were prepared according to the data collection and solving difficulties and it was issued for both chemistry teachers and chemistry students. After the data had been collected from the questionnaires, they were analyzed. Then the problems were identified and suggested modifications were recommended.

Analysis of the answers of questionnaire for students indicated various misconceptions and naive ideas that most of the students have in this topic. As an example, a question of 'What do you mean by the resonance phenomena"

It was obvious that some students could not answer while some others are interpreted the questions in a wrong way and some of them had given completed answers.

These results suggested that the students liked the structure of the alternative teaching method following by a graded sequence in small steps so that students can work according to their own learning speed, having no time restrictions, and being able to use slides which were given to them as a supplementary material at home. All these views are in agreement with the previous studies. Since this alternative teaching approach is cumulative, attendance at the course is one of the most important factors affecting the achievement.

Here the same post – test being administered to both groups, even though, the performances were different. Experimental group performed well than the controlled group.

Chemistry teachers must make much effort to create an ideal environment for teaching and learning by including technological tools in the classroom will require teachers to employ different teaching techniques.

Interactive computer-generated molecular visualizations of the resonance phenomena provide many opportunities for instructional innovation. They are being used not only to encourage students to think about chemistry in terms of resonance contributors, resonance hybrid, and electrons symbols, but also to provide opportunities for students to become more independent learners. Further, this interactive visualization can help teachers' interest and motivate students with a variety of learning styles. To achieve these goals, more activities incorporating visualizations that are easy for students to use and interpret will need to be developed.

4.2 Conclusion

There is a significance difference in the pre and post tests for the experimental group, as an alternative study, technical tools and visualized aids can be successfully used in the enhancing resonance studies among the G.C.E.A/L Chemistry students.

Further, this statement is powered, because mean and highest marks in the post-test for the experimental groups are high and the minimal marks are low for the same. Most of the teachers and students are satisfied with the introduced method. Technological tools are beneficial and effective in the development of new methods and techniques in the designing and implementation of alternative teaching methods. Outcomes of the students regarding the resonance have been improved by using the particular, alternate teaching method.

4.3 Recommendations

Instead of the conventional teaching method, teacher has to adopt various other ways of innovative teaching methodologies with the help of audio- visual teaching aids and computer assisted teaching. For example- preparation of PowerPoint slides for each topic, and using educational Compact disks. Teachers should be given opportunity to follow the course in information technology so as to enhance their knowledge in using computers and preparing study packages for educational purposes. Using both pre-test and post-test results, it was shown statistically that while there was no difference between the groups prior to intervention, the experimental group performed significantly better than the control group after the treatment. Data analyses revealed the mean post test scores of the experimental group is higher than the mean post test scores of the controlled group.

Moreover, experimental group students showed best understanding of the concepts in 42 % in post test. Bar – chart representation of the student's competency level of post test marks is also supportive of the comparatively better performance of experimental group since; the number of students in the higher competence level show considerable increase in the experimental group compared with controlled group.

This is consistent with the claims made for alternative teaching approach based on student impression. The limited but objective results in this research suggest that substantially self-paced alternative learning approach is a better technique than the conventional teaching in the resonance topic. Another important aspect is that this alternative teaching approach forces student active participation in the teaching-learning process. It shifts the responsibility for learning back to the student, where it should be. Because it provides for a self-paced, logical sequence of small steps, and immediate confirmation or correction, it helps to overcome the wide spread of abilities and interest among A/L chemistry students.

Another positive effect of the intervention on the concept understanding of students in chemistry stemmed from the creation and display of slides with animation diagrams to the experimental group in the activities. The opportunities for the experimental group communicating their knowledge of chemistry through the visual techniques with the animation diagrams contribute to understand the "Resonance" topic in chemistry. The alternative teaching strategies positively affected students' conceptual understanding in chemistry and it was useful to enhance their interest to learn subject matter of the chemistry and develop their active participation.

However, a problem with the using of animations for teaching molecular structure and dynamics in "Resonance" is that merely viewing visualization may lead to learning at a lower level than would draw or build a molecular structure. Consequently, it may be beneficial for computer-based visualizations to provide opportunities for students actively to explore concepts.

References

1. Ardac, D., & Akaygun, S. (2004). Effectiveness of multimedia-based instruction emphasizes molecular that representations on students' understanding of chemical

   change. *Journal of Research in Science Teaching*, 317-337.

2. Clark, D., & Jorde, D. (2004). Helping students revise disruptive experientially supported ideas about thermodynamics: Computer visualizations and tactile models. *Journal of Research in Science Teaching*, *41*(1), 1-23.

3. Coll, R. K., & Treagust, D. F. (2003). Investigation of secondary school, undergraduate, and graduate learners' mental models of ionic bonding. *Journal of Research in Science Teaching*, *40*(5), 464-486.

4. Daraniyagala, S.P.(2011). *Journal of Chemistry in Sri Lanka*. Vol 25(3), Vol 26(2)

5. Gabel, D. L., Samuel, K. V., & Hunn, D. (1987). Understanding the particulate nature of matter. *Journal of Chemical Education*,64(8), 695-697.

6. Gopalan, R. (2009) Inorganic Chemistry for Undergraduates. University press India private limited 206-212.

7. Gurdeep, R., (2008) Advanced Inorganic Chemistry. thirty first edition, Krishna prakashan media(P) Ltd,India 145-155.

8. Haidar, A. H. (1997). Prospective chemistry teachers' conceptions of the conservation of matter and related concepts. Journal of Research in Science Teaching,34(2), 181-197.

9. Holleman, A. F., Egon Wiberg, & Nils Wiberg. (2001), Inorganic Chemistry. Academic press, USA, 125-130.

10.Johnson-Laird, P. (1983). *Mental models*. Cambridge: Cambridge University Press.

11.Kozma, R. B., & Russell, J. (1997). Multimedia and understanding: Expert and novice responses to different representations of chemical phenomena. Journal of Research in Science Teaching, 34(9), 949-968.

12. Mayer, R. E. (2003). The promise of multimedia learning: Using the same instructional design methods across different media. *Learning and Instruction*, *13*(2), 125-139.

13. Othman, J., Treagust, D. F., & Chandrasegaran, A. L. (2008). An investigation into the relationship between students' conceptions of the particulate nature of matter and their understanding of chemical bonding. International Journal of Science Education, 30(11), 1531-1550.

14. Petrucci, Harwood, & Herrine. Principles and modern application. Prentice Hall Upper saddle river. Eighth edition 460 – 464.

15. Philip Matthews,. (2002). Advanced Chemistry. Cambridge University press, 78 – 83p.

16. Raymondchang, (2010). Chemistry. Tata McGraw Hill Education. Ninth edition. 372 – 383.

17. Ross, B., & Munby, H. (1991). Concept mapping and misconceptions: A study of high-school students' understandings of acids and bases. International Journal of Science Education,13(1), 11-23.

18.Sanger, M. J., Phelps, A. J., & Fienhold, J. (2000). Using a computer animation to improve students' conceptual understanding of a can-crushing demonstration. Journal of Chemical Education,77 (11),1517-1520.

19. Sellke., Thomas., Bayarri., Berger M. J., James O. (2001). Calibration of p Values for Testing Precise Null Hypotheses. *The American Statistician* 55 (1), 62–71.

20. Singh. D.N. (2010) Basic Concepts Of Inorganic Chemistry Dorling Kindersley (india) Pvt .Ltd 62-67

21. Yuli-Rahmawati (2008). The Role of Constructivism in Teaching and Learning Chemistry, Retrieved January 16, 2013, from

    http://pendidikansains.wordpress.com/2008/04/14/the-role-of-constructivism-in-teaching-and-learning-chemistry/

APPENDICES

Appendix 1 - Questionnaire for A/L chemistry teachers

This questionnaire is provided to you to get information for my research study under the title 'An alternative educational approach for resonance in advance level chemistry'. I do hereby assure, all the information provided by you will be treated as strictly confidential and used solely for the purpose of this research project. No individual respondents will be identified in the report. I humbly request your kind cooperation for the successful completion of my research.

- G.Nakkeeran

This is a questionnaire provided you to get some ideas about teaching resonance in A/L chemistry.

1 General information

    1.1 Name of the school..

    1.2 Type of the school..

    1.3 Gender ....

    1.4 Educational Qualification...

    1.5 Period of Service in teaching ....

    1.6 Number of years served as a Chemistry teacher.......................................

- **Put a tick (✓) in the appropriate cage**

2 Do you think that the resonance is the important topic in A/L chemistry?

                        Yes ☐

                        No ☐

3 If your answer to question 2 is No, what are the reasons?

    Less important in exam point of view   ☐

It covers only a very little area in chemistry ☐

It is not connected with any other topic in chemistry ☐

Other reasons (Please specify the reasons) ....

4 Do you think that the introduction of animation diagrams to show the particular resonance phenomena will have favorable impact on understanding the resonance topic?

Yes ☐

No ☐

5 If your answer to question 4 is No, what are the reasons?

No need to explain this topic with animation diagram ☐

Animation diagrams are difficult to understand ☐

Students won't give their full attention on animation diagrams ☐

Other reasons (Please specify the reasons) ....

6 Do you think the need of alternative teaching material is important to teach the resonance topic?

Yes ☐

No ☐

7 Do you use computer facilities to teach the topic – 'Resonance' in your class?

Yes ☐

No ☐

8 Do you think the power point based presentation with the use of Over Head Projector (OHP) is important to teach the resonance topic?

Yes ☐

No ☐

9 Are you able to teach this particular topic in your school with the resources available?

$$\text{Yes} \quad \boxed{}$$
$$\text{No} \quad \boxed{}$$

10 What do you think as the major drawback in your school to present the topic – "Resonance" in a more effective manner?

Lack of resources $\quad\boxed{}$

Allocated time period is not enough $\quad\boxed{}$

Poor interest of students $\quad\boxed{}$

Other reasons (Please specify the reasons) ……………………………………..

11 Your suggestions about teaching Resonance in an effective manner?

……………………………………………………………………………………

……………………………………………………………………………………

……………………………………………………………………………………

……………………………………………………………………………………

……………………………………………………………………………………

……………………………………………………………………………………

Appendix 2 – Pre Test Questionnaire

1. Which of the following molecule / ion has, no resonance phenomena?

   (i) O_3 (ii) NO_2 (iii) SO_4^{2-} (iv) ClO_3^- (v) NH_4^+

2. The number of possible resonance structure of MnO_4^- is?

   (i) 3   (ii) 4   (iii) 5   (iv) 6   (v) 8

3. Which of the following species as the resonance phenomena?

   (i) $CH_2=CH-CH_2-CH_2-CH=CH_2$
   (ii) $CH_2=CH-CH_2-CH_2=CH-CH_2$
   (iii) $CH_2=CH-CH_2=CH-CH_2-CH_3$
   (iv) $CH_2=CH-CH_2-CH_2-CH_2^+$
   (v) $CH_2=CH-CH_2-CH^+-CH_3$

4. Which one of the followings statement is not true?

   i)   The net charge of the species is conserved in the resonance phenomena.
   ii) When no of resonance structures increases stability of species increase.
   iii) The bond strength increases to the increase of bond order
   iv) All resonance structures of ion or molecule have the same energy
   v)   The two O – O bonds of ozone are bond equal in length

5. Which one of the followings closely resembles the actual structure of SO_4^{2-} ion

   (i)   (ii)   (iii)   (iv)   (v)

52

6. The actual structure of NO_2F is

(i) $:\overset{\cdot\cdot}{\underset{}{O}}$ $:\overset{\cdot\cdot}{F}-\overset{\parallel}{N}=\overset{\cdot\cdot}{O}\cdot$

(ii) $:\overset{\cdot\cdot}{O}:^-$ $:\overset{\cdot\cdot}{F}=\overset{+}{\underset{+}{N}}-\overset{\cdot\cdot}{O}:$

(iii) $:\overset{\cdot\cdot}{O}:^-$ $:\overset{\cdot\cdot}{F}-\overset{2+}{N}-\overset{\cdot\cdot}{O}:^-$

(iv) $:\overset{\cdot\cdot}{O}:$ $:\overset{\cdot\cdot}{F}-\overset{\uparrow}{N}=O:$

(v) $:\overset{\cdot\cdot}{O}:^-$ $:\overset{\cdot\cdot}{F}-N\equiv\overset{\cdot\cdot}{O}^+$

(6 × 5 = 30 Marks)

7. Draw the Lewis structure for the following ions/ molecules?

i) H_2O_2 ii) CO_2 iii) CN^- iv) OH^- v) HCl

....

....

....

....

....

....

....

(5 Marks)

8. Draw the Lewis structure for Ozone (O_3)

....

....

(1 Marks)

9. What do you mean by the resonance phenomena?

....

....

....

(10 Marks)

10. Draw the all, possible resonance structures for CO_3^{2-} ion?

....

....

....

....

(10 Marks)

11. Give the bond order for the following species?

i) H_2 ii) O_2 iii) N_2 iv) O_3 v) CO_3^{2-}

....

....

....

....

....

(10 Marks)

12. Give the bond order of $Cl - Cl$ bond?

....

....

(2 Marks)

13. The stability of $CH_2 = CH - CH_2^-$ is greater than that of

$CH_2 = CH - CH_2 - CH_2^-$ explain?

....

....

....

....

....

(10 Marks)

14. The part a and b base the regarding with the structure of HCO_3^-. The skeletal structure of HCO_3^- is given below

$$H - O - C \overset{\overset{O}{|}}{} - O$$

a) Draw the Lewis structure of HCO_3^-

....

....

....

(1 Marks)

b) Draw the all possible resonance structure of HCO_3^- and comment on the stability of resonance structure.

....

....

....

....

(10 Marks)

15. i) draw the Lewis structure of azide ion (N_3^-)

....

....

....

(1 Marks)

ii) Give the possible resonance structures of azide ion?

....

....

....

....

....

....

(10 Marks)

Appendix 3 – Post-test Questionnaire

1. Consider the oxi anions of chlorine ClO^-, ClO_2^-, ClO_3^- ClO_4^-

   Which of the following statement is not true regarding with these oxi anions

   a) No of resonance structures increases in the order of ClO^-, ClO_2^-, ClO_3^- ClO_4^-

   b) Chlorine atoms of all oxi anions are in SP^3 hybridization

   c) The stability of oxi anions increases order of ClO^-, ClO_2^-, ClO_3^- ClO_4^-

   d) The bond length decreases order of ClO^-, ClO_2^-, ClO_3^- ClO_4^-

   e) Bond order decreases in the oxi anions of ClO^-, ClO_2^-, ClO_3^- ClO_4^-

2. Which one of the following pairs does not represent a resonance structure?

   1)

   2)

   3)

   4)

   5)

3. Which one of the followings closely resembles the actual structure of SO_4^{2-} ion

i)
$$O=S=O$$ with $2-$ and O below, O above

ii) $$\left[\begin{array}{c} O \\ O-S-O \\ O \end{array} \right]^{2-}$$

iii) $$O=S-O^- \text{ with } O^- \text{ below}$$

iv) $$-O-S-O^- \text{ with } O \text{ above and } O^- \text{ below}$$

v) $$-O-S-O^- \text{ with } 2+, O \text{ above and } O \text{ below}$$

4. The actual structure of NO_2F is

i) $$\ddot{F}-\ddot{N}=\ddot{O}\cdot$$ with Ö above

ii) $$\ddot{F}=\overset{+}{N}-\ddot{O}:^- $$ with :Ö: above

iii) $$\ddot{F}-\overset{2+}{N}-\ddot{O}:^- $$ with :Ö:$^-$ above

iv) $$\ddot{F}-\ddot{N}=\ddot{O}:$$ with :Ö: above

v) $$\ddot{F}-N\equiv\overset{+}{O}$$ with :Ö:$^-$ above

5. Correct decreasing order of bond length of N-O in NH_2OH, NO, NO_2^-, NO_3^-

1) $NO_2^- > NO_3^- > NO > NH_2OH$

2) $NO_3^- > NO_2^- > NO > NH_2OH$

3) $NO > NO_2^- > NO_3^- > NH_2OH$

4) $NH_2OH > NO_3^- > NO_2^- > NO$

5) $NO > NO_3^- > NO_2^- > NH_2OH$

6. Draw the Lewis structures of HSO_4^-

....

....

....

....

58

7. Draw the all, possible resonance structures of HSO_4^-

....

....

....

....

8. Give the bond order of S – O species

....

....

9. a) Draw the Lewis dot structures and the resonance structures of HNO_3

....

....

b) Give the all resonance structure for HNO_3

....

....

....

....

....

....

$$O$$
$$\uparrow$$
$$H - O \xrightarrow{a} N \overset{b}{\underset{c}{=}} O$$

c) Give the relationship between bond length of a, b and c ?

....

....

....

....

10. The skeletal structure of per ox nitrate ion/ OONO is given bellow

$$O - O - N - O$$

a) Draw the acceptable Lewis structure of per oxo nitrate ion.

....

....

b) give all the possible resonance structures of above ion.

....

....

....

11. Arrange the following molecules in the increasing order of O – O bond length

a) H_2O_2 b) O_3 c) O_2

....

....

12. i) draw the Lewis structure of azide ion (N_3^-)

....

....

ii) Give the possible resonance structures of azide ion?

....

....

....

....

13. Arrange the following species in the increasing order of bond length

a) CO   b) CO_3^{2-} c) CO_2

....

14. Give the Lewis structure of HNO_3, $H_2 S_2O_7$, $S_2O_3^{2-}$, HN_3, ClO_4^-

....

....

....

....

....

15. The skeletal structure of methyl isocyanate is given below.

$$H-\overset{\overset{\displaystyle H}{|}}{\underset{\underset{\displaystyle H}{|}}{C}}_{(1)}-N-\overset{}{C}_{(2)}-O$$

a) Draw the appropriate Lewis structure?

....

....

b) Draw the all possible resonance structure of isocyanate ?

....

....

....

c) State the stability of them?

....

....

....

Appendix 7

Lesson plan – 1

Grades	– 12 and 13 (together)
Date	– 12.02.2013
Subject	– Chemistry
Unit	– 2
No. of periods	– 2
Competency	– Use electronic arrangements in determining molecules
Quality inputs	– Leaflets, Pictures,

Learning and teaching process –

Steps:-

1) Divide the students into groups

2) Ask about the characteristics of electrons

3) Give brief instruction about electrons

4) Give instruction about atoms

5) Introduce the Lewis structures for atoms

6) Draw the Lewis structures for the given atoms

7) Ask to comment on ionization of elements

8) Clarify the phenomena "electro negativity"

9) Guide the students to report their collected information in different ways such as oral presentation and written reports.

Evaluation – Evaluate the students by their presentations and reports.

Feedback – At the end of the students' presentation, students were praised and encouraged to do in a better way next time and also information gap in the presentation was filled. Students' reports were corrected and constructive comments were written.

Appendix 8

Lesson plan – 2

Grades	– 12 and 13 (together)
Date	– 13.02.2013.
Subject	– Chemistry
Unit	– 2
No. of periods	– 2
Competency	– Uses electronic arrangements in determining molecules
Quality inputs	–

Learning and teaching process –

Steps:-

1) Divide the students into groups
2) Ask to comment on ionization of elements
3) Clarify the phenomena "electro negativity"
4) Ask about the formation of covalent bonds
5) Explain about the requirements for the formation of covalent bonds
6) Introduce the two types of covalent bond σ and π
7) Briefly introduce the Lewis structure for covalent bond
8) List the fundamental requirements for the formation of Lewis structure
9) Describe the underlying theories for drawing Lewis structure
10) Illustrate the Lewis structures for given covalent molecules
11) Write the numbers of σ and π bond in each, drawn molecule.
12) Guide the students to report their collected information in different ways such as oral presentation and written report.

Evaluation – Evaluate the students by their oral answers, written answers and drawings

Feedback – Faults and mistakes of students were identified and corrected. Students were encouraged to do best next time.

Lesson plan – 3

Grades	– 12 and 13 (together)
Date	– 19.02.2013
Subject	– Chemistry
Unit	– 2
No. of periods	– 3
Competency	– Uses electronic arrangements in determining molecules
Quality inputs	– Colored pictures, Painting sticks, blue tack

Learning and teaching process –

Steps:-

1) Divide the students into groups

2) Ask to comment on the phenomena "resonance"

3) Explain the underlying theories on resonance

4) Clarify the relationship of bond order, bond length and bond strength regarding the stability of covalent bonds

5) List four instances where the resonance occur

6) Draw the resonance structures for given molecules and ions

7) Calculate the bond orders for the given molecules and ions

Evaluation – Evaluate the students by their presentations and reports.

Feedback – Faults and mistakes of students were identified and corrected. Students were encouraged to do best next time.

Appendix 10

Lesson plan – 4

Grades	– 12 and 13 (together)
Date	– 20.02.2013
Subject	– Chemistry
Unit	– 2
No. of periods	– 3
Competency	– Uses electronic arrangements in determining molecules
Quality inputs	– A4 sheets, pen, pencil, painting sticks

Learning and teaching process –

Steps:-

1) Divide the students into groups

2) Ask about the resonance

3) Draw the resonance structures for given molecules

4) Indicate how many resonance contributors are for each molecule

5) Compare the stability of each resonance contributor for each molecule

6) Write the reasons for the stability and instability of the abovementioned resonance contributors

7) Calculate the bond order for each molecule given in question 3 and predict the stabilities according to the bond order

Evaluation – Evaluate the students by their presentations, reports and drawings.

Feedback – Faults and mistakes of students were identified and corrected. Students were encouraged to do best next time.

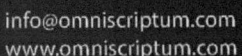